METAL DETECTING FOR TREASURE

A Guidebook for Beginners

by Dorothy B. Francis

Published by
BALLYhoo BOOKS
P.O. Box 534
Shoreham, NY 11786

DEDICATION

This book is for those just beginning to realize the excitement and mystery of searching for treasures lost or hidden. May the ground be damp, the sand dry, and the water calm.

METAL DETECTING FOR TREASURE
A Guidebook for Beginners

Text: Copyright © 1992 by Dorothy B. Francis

ALL RIGHTS RESERVED. No part of this book may be reproduced or transmitted in any form or by any means, electronic or mechanical, including photocopying, recording or by any informational storage or retrieval system — except by a reviewer who may quote passages in a review to be printed in a magazine or newspaper — without permission in writing from the publisher. For information contact BALLYHOO BOOKS, P.O. Box 534, Shoreham, NY 11786.

First Printing: May, 1992

ISBN: 0-936335-03-3

Library of Congress Cataloging-in-Publication Data

```
Francis, Dorothy Brenner.
  Metal detecting for treasure : a guidebook for beginners / by
Dorothy B. Francis.
      p.     cm.
  Includes bibliographical references.
  Summary: Explains how to use metal detectors to locate and
retrieve coins, jewelry, or gold. Includes anecdotes of people who
"struck it rich."
  ISBN 0-936335-03-3 (pbk.) : $8.00
  1. Treasure trove--Juvenile literature.  2. Metal detectors-
-Juvenile literature.   [1. Buried treasure.  2. Metal detectors.]
I. Title.
G525.F63   1992
622'.19--dc20                                         92-12506
                                                          CIP
                                                           AC
```

Thanks to: Shelley Lantheaume of Plaza Graphic Associates
           Robert Parisi of R.J.P. Graphics
Book design: Brian J. Heinz

Printed in the United States of America   **ISBN: 0-936335-03-3**

ACKNOWLEDGEMENTS

I would like to express my sincere thanks to the many park commissioners, police officers, grounds caretakers, and private property owners who allowed me to use my metal detector on land under their jurisdiction while gathering material for this book.

A special thanks goes to my husband, Richard Francis, for his help with photography.

I would also like to thank Jim Lewellen, president and general manager of the Fisher Research Laboratory, for his willingness to assist me in locating factual material and for supplying photographs.

Readers may write to the following address for more information about metal detectors:

Fisher Research Laboratory
200 W. Willmott Road
Dept. MDT
Los Banos, CA 93635
Ph. 209-826-3292   Fax 209-826-0416

TABLE OF CONTENTS

Introduction 1

Dr. Gerhard Fisher — A Brief Remembrance 3

The Lure and Lore of Treasure 5

Choosing Your Treasure Hunting Equipment 11

Read Your Manual 16

Retrieving Your Treasure 22

Treasure Awaits You 26

Treasure on Land 31

Treasure at the Beach 36

Treasure in the Surf 42

Prospecting for Treasure 45

Safety Tips 48

Finders Keepers 49

Bibliography and Resources 51

Introduction

As a child I remember being bored with WINNIE THE POOH, and begging the teacher to read TREASURE ISLAND. I liked the "Uncle Wiggly" tales, but I preferred hearing my grandfather tell pirate stories, accounts of the great hurricane that sank treasure-laden ships at Galveston Island in the early 1900's, and stories of people burying their life savings under a fencepost during the depression. The idea of hidden treasure fascinated me.

Then, in the 1970's, I met Key West treasure salvor, Mel Fisher, who was searching for the Spanish galleon, *Atocha*. Aboard his replica galleon he showed me doubloons, pieces of eight, and gold chains he had retrieved from the sea. I felt the weight of gold bars as he told of the hypocritical and marauding Spaniards who stamped the bars with both the mark of the conquistadors and the Christian cross. I decided then that one day I would find a gold coin from Old Spain.

I'm still looking for it.

In 1989 I bought a very inexpensive metal detector and took it to the Keys. I'm convinced that there's a master plan that brings people together when the time is right. And that's why I met those two treasure hunters while I was beach searching for that gold doubloon.

One man headed directly <u>into the surf</u>. I was wide-eyed. I had never heard of taking a <u>metal detector</u> into the water.

"He's using a Fisher 1280-X Aquanaut," the other man said, answering my unasked question. "He can take the whole thing right into the surf where the good stuff is."

This man paused to talk to me after retrieving coins from a volley ball court that I had just searched, finding nothing.

"Why didn't I find those?" I asked.

"Your detector's only good for hanging on the wall." He laughed. "Get a decent machine if you expect to find anything."

He was using a Fisher 1235-X. I memorized the name. That afternoon I looked up the 800 numbers of detector dealers in a treasure magazine, called several, and discussed the 1235-X. When one dealer asked where I had learned about it, I told him about the treasure hunter on the beach. After a long silence, the dealer replied.

"That's like buying a Caddie because someone tells you it's better than your Ford." But he agreed that the Fisher 1235-X was right for beginning or advanced hobbyists. I ordered it. This is a poor way to choose a metal detector. I had lucked out. That anonymous treasure hunter probably won't be reading a book for beginners, but if he should, I'd like to say, "Hey, thanks!"

Dr. Gerhard Fisher — A Brief Remembrance

Dr. Gerhard Fisher has often been called the Henry Ford of metal detectors. He came to America in 1923, holding a degree in electrical engineering from the University of Dresden, Germany. He worked in New Jersey researching radio circuitry before moving to Los Angeles where he secured a position as a research engineer for Western Air Express. His original experiments concerned the development of aircraft radio direction finders.

Pilots using these instruments reported errors in the location of signals when they flew over metal or heavily mineralized soil. These reports gave Dr. Fisher the idea of developing and using the 'null' system to locate underground metal, and in his spare time he developed an instrument that would perform this task. Being a man of vision, he could see many potential uses for such an instrument.

In 1931 Dr. Fisher founded Fisher Research Laboratory in his Palo Alto, California garage, and in 1937 he was granted the first metal detector patent. His work in radio circuitry attracted the interest of Dr. Albert Einstein. Although Dr. Einstein correctly predicted the world-wide use of radio direction finders in the air, on land, and at sea, Dr. Fisher proved erroneous Einstein's belief that metal detectors would be of little interest or use to anyone.

Mr. Jim Lewellen, today's president and general manager of the Fisher Research Laboratory, has the following to say about Dr. Gerhard Fisher: "Not only was Dr. Fisher a brilliant scientist and entrepreneur, he was also a man of vision and principle. And he had backbone, gumption, and determination to back up his principles and transform his vision into reality."

"Today the world is richer for Dr. Gerhard Fisher's contributions."

Courtesy: Fisher Research Laboratory

Dr. Gerhard Fisher, 1899-1988, founder of Fisher Research Laboratory and the recipient of the first metal detector patent in 1937, began making metal detectors in his garage in 1931.

The Lure and Lore of Treasure

People from all walks of life are fascinated by the lure of things lost or hidden. A child spotting the gleam of a quarter under the football bleachers knows the thrill of discovery. The adult running his searchcoil over a long-forgotten park and hearing a smooth audio signal flow though his headphones knows the same thrill as the child. The prospector searching worked-out mining country experiences similar exciting sensations when his metal detector whispers of gold. Perhaps the search for riches meets some primitive need built into human genes.

Who doesn't experience a feeling of awe as he reads about Wallace Chandler's Chesapeake Bay discovery of a man's natural sapphire ring that has been appraised for more than $25,000?

"The sound coming from my 1280-X headphones was much like that of a fishing sinker, Chandler said, "but I carefully lifted the target to the surface. Zounds! Eureka! Bingo! Or whatever one says when he hits the jackpot! I picked a man's gold ring from the scoop. And what a ring it was! The high-karat gold shone almost as bright as the day it was lost — decades, perhaps even a century ago. The large blue gemstone glowed with exceptional color and beauty."

A few more people probably bought water detectors and rushed into the surf after reading of Wallace Chandler's experience.

Who can resist reading about Mel Fisher and his quest for and fabulous discovery of the *Atocha* and its sister galleons?

Also in the Florida Keys one hears whispers about 'that man who has found more gold rings than he has fingers for.'

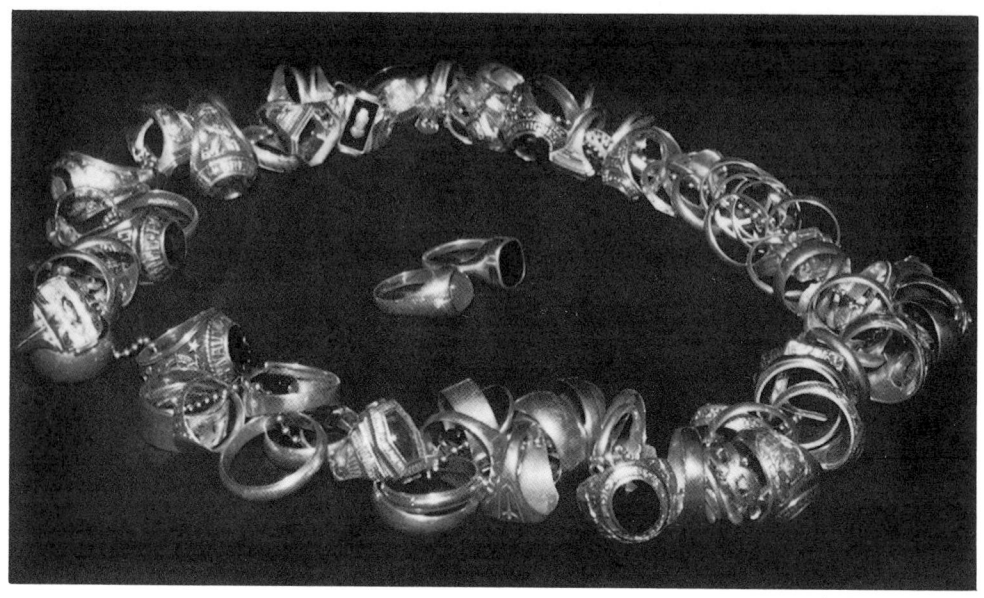

Michigan treasure hunter, Wallace Chandler, spends most of his spare time metal detecting in shallow water. These are just a few of the many gold rings he's found. The antique blue sapphire ring (right-center above) has been appraised at $25,000.

If you don't hear enough word-of-mouth tales to make you want to try metal detecting, there are several magazines devoted to the hobby which will intrigue you. They have no trouble keeping their pages filled with stories of newly discovered caches of gold and silver, intriguing Civil and Revolutionary War artifacts, and collector's specimens of antique bottles, buttons, and coins.

The doorway to the hobby and pastime of treasure hunting with a metal detector is open to anyone who cares to enter. When I attended the H.E.A.R.T. organized hunt in Waverly, Iowa in 1991, I was amazed at the diversity of the people in attendance. (H.E.A.R.T. is the Historical Educational Archaeological Research Team.)

About one tenth of the participants were women. There were husband and wife teams and there was one family — husband, wife, and son. I met a teacher, a librarian, a postal worker, a school custodian, a writer, a telephone executive, a retired railroad executive, a cement worker, and an electrician. All these people were drawn together for the enjoyment of their hobby.

Many people are curious about metal detectors and wish they had one, yet they shy away, believing the expense will be too great. Not true. There are fine, yet relatively inexpensive, detectors available that are suitable for the beginner or for the whole family. A metal detector may be less costly than quality golf clubs, a fine fly rod, or a good shotgun.

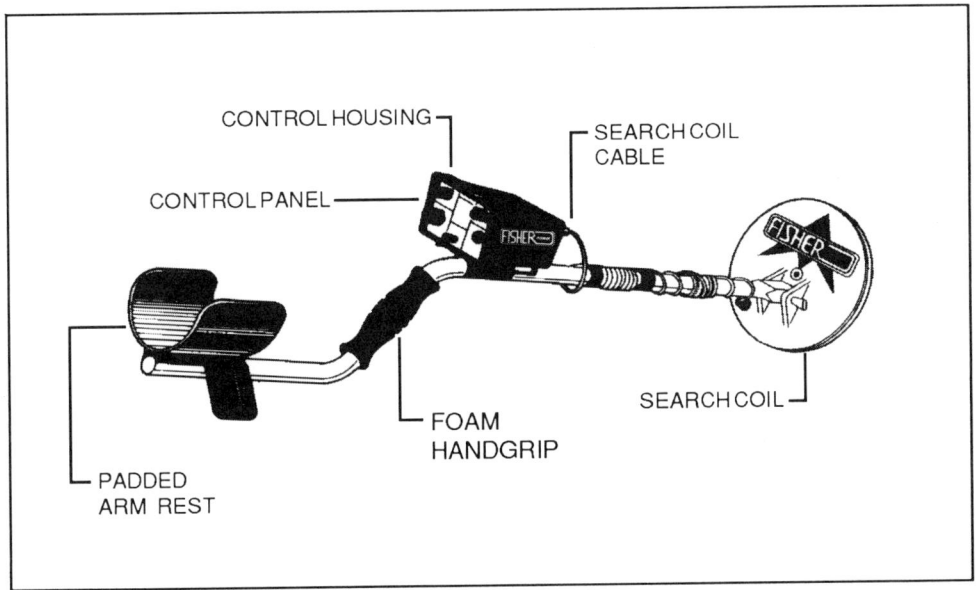

Many of today's modern metal detectors are lightweight, comfortable and easy to use. This Fisher M-Scope Model 1235-X features a "convertible" control housing which slips off the stem for belt or chest mounting. Smaller and larger searchcoils are also available.

A family affair? One detector for several people, you may ask? It may be more fun to have a detector for each family member, but for starters, a single detector will allow one person to mark the targets, another person or persons to help dig those targets, and yet another to keep a log of the finds. Once a family becomes deeply involved in the hobby, more detectors can be added as needed.

Few people expect their hobbies to pay for themselves, but metal detecting is one leisure activity that may do this. A hobbyist who uses his equipment with regularity may pay for that equipment with the coins, jewelry, and artifacts he finds.

The treasures one discovers through metal detecting may be far more important and fascinating to the finder than monetary riches. Lucile Bowen, a treasure hunter for over 25 years and a frequent contributer of articles to "Eastern and Western Treasures" magazine has found nuggets, diamond rings, gold pieces, and more thousands of coins than she has kept track of.

But of all her finds, one of her favorites is an old locket with a Turkish symbol on the front. Inside the locket is a tiny gold heart with little teeth marks on it. In olden times people used to rub the babies' gums with gold or silver to help the teeth work their way through. In studying the locket Lucy must wonder, what baby? How long ago? Did it survive?

Lucy also prizes a stick pin showing a deer jumping over a hand plow. The reverse bears the words *John Deere*. The locket and the stick pin are probably worth far less than Lucy's other finds, but they carry great sentimental value.

Family togetherness is another priceless treasure. Parents who introduce kids to metal detecting may be supplying them with a lifetime hobby and one that will help see them through some growing-up years when peer pressure might tempt them into less wholesome pursuits. When night comes, a teen who has spent the day digging for treasure may be more eager to head for bed than for the local bar.

Young people may care little for history until they find an old coin and seek information about it. One boy began a Lincoln cent collection after finding a wheat-back penny with the designer's initials on it. He subsequently learned that the Lincoln cent first issued in 1909 marked the centennial of Lincoln's birth. When the likeness of the Lincoln memorial replaced the wheat design on the reverse of the coin in 1959, he knew that it marked the 150th anniversary of the great man's birth. He also became interested in the Indian head pennies that preceded the Lincoln head cents. A sense of history and the richness it adds to one's life can evolve through treasure hunting.

In this age of stress, pressure, and career burnout, treasure hunting can be the answer when the doctor says, 'find a relaxing hobby that will get you outdoors and let you forget the woes of the workplace.' Metal detecting can offer inexpensive hours of therapy that may help prevent a serious illness.

Some detectorists have become Civil War buffs after finding a fascinating artifact. One woman I talked with said that searching for artifacts held little appeal. But after finding old musket balls on her Kansas property, she changed her mind. Awed by the significance of her finds as they related to Kansas-Missouri border wars and the raids by Quantrill and his riders, she became hooked on artifacts and their history.

Treasure hunting can lead to other spin-off interests. Many coin collectors became involved in that hobby through studying the coins their detectors targeted. The same thing has happened to bottle collectors who have located bottles through their metal caps or lids, and many buttons and key collections began with a metal detector find.

People who are lonely may want to consider taking up treasure hunting. Metal detectors attract the curious. I can't remember a day of metal detecting in which someone didn't approach me with questions or comments. Some of them were long-time treasure hunters who wanted to share their experiences. Others just wanted to know what I was doing, if the machine worked, what I was finding. A person with a metal detector will have a plethora of people to talk to.

An interesting way to expand your enjoyment of treasure hunting is to join a metal detector club. Your membership will strenghthen the organization as it enables you to learn more about your detector, to meet new friends, to join in organized field trips and competition hunts, and to share your stories of discovery with others.

A good group to contact is:

**The Federation of Metal Detector & Archaeological Clubs, Inc.
1614-0 Union Valley Road - Suite 131
West Milford, NJ 07480-2222**

This group was established in 1984. Its mission is to unite, promote, and encourage metal detecting clubs; to spread information; and to take action to preserve and protect the hobby. If you write to this group, they will send you the name of the FMDAC club nearest you. From that information you'll be able to locate other area clubs.

The doorway to pleasure and treasure is open. Why not enter? You'll be welcome.

Choosing Your Treasure Hunting Equipment

The metal detector is the workhorse of the treasure hunter, so choose this tool carefully. Your choice of detector will depend on the type of treasure you're seeking.

You may plan to search for coins, for large caches of deeply buried loot, for artifacts, for valuables lost in water, or for gold. There are metal detectors especially made for finding the type of treasure you seek. Also, there are all-purpose detectors that can be used for several kinds of treasure hunting.

For the beginning hobbyist, unless you plan to prospect for gold, a multi-purpose, turn-on-and-go detector will probably be the most desirable. Since gold is sometimes found in minute amounts and in highly mineralized ground, prospectors should buy a detector with circuitry especially designed for finding gold such as Fisher's Gold Bug.

Unless you're planning to prospect for gold, be sure your detector has the ability to discriminate, that is to reject foil, bottle caps, pull-tabs, etc. Once you're in the hunt field, you can then decide what amount of discrimation you want to use. Gold prospectors usually prefer to discriminate nothing.

Some people prefer to buy top-of-the-line equipment when beginning a new venture. Others choose to start with less expensive instruments, upgrading as the extent of their interest increases. It's usually sound advice to purchase the best detector that you can afford or care to afford.

There are several good brands on the market, but I'm pleased with my Fisher 1235-X because it does a great job of targeting coins and jewelry. I also like the convertible control housing that allows me to mount it on a belt or a neck strap, thus allowing shallow water searching. The 1235-X is easy to use. With only three knobs and a pinpoint button I have no need to tinker with a lot of gadgets in order to get the instrument turned or ground balanced. I snap it on and go.

To learn about the detectors available, talk with others who pursue this hobby, getting their opinions, then read treasure magazines, their ads, their editorial columns, their articles. Give special attention to field-test reports. Some reports may not be meaningful to you, but don't let the 'detectorese' scare you. The basic concepts will seep into your thinking.

Read any books you can find about treasure hunting. The treasure magazines usually advertise some volumes that will interest you whether or not you decide to become an active participant in the hobby.

After you've done some reading, call on local dealers, and see what they have to offer. Write the detector companies for more information on their products. Call the 800 numbers you've found in detector advertisements and talk with those dealers, asking for answers to questions that puzzle you.

Finding a local dealer may pose a problem. Look in the telephone directory's yellow pages. If there's a dealer listed in your community, by all means call on him and look over the detectors he has in stock. Ask for demonstrations. Buying from a local dealer gives you the advantage of being able to use your machine in the field, then return to the dealer for more advice if you need it. It's good to have this kind of backup.

If there is no dealer in your area, call the detector company or call the 800 numbers listed by the mail-order dealers. Compare prices. Note the helpfulness in the voices at the other end of the wire. In making your purchase, try to choose a dealer who operates a retail store as well as a mail order business. His shop may be several hundred miles away, but the day may come when you'll have the opportunity to stop in and talk with him personally. Most dealers are also active detectorists and they have a fund of information to offer the beginner.

Once you've decided on the detector of your choice, plan to buy headphones for it. You may even want to buy two sets — an 'earmuff' type for cool weather and a lighter type to use in summer heat. Headphones serve several purposes. They enable you to hear detector signals more clearly. Listen to a signal without using headphones. Then go over the same signal with headphones in place. Note how much more clear the signal is. Headphones help minimize the distractions of people talking, wind, traffic, surf sounds. You'll also find that using headphones will prolong battery life.

The use of headphones helps insure your privacy. Unless an observer is standing very close to you, he'll be unable to hear the detector signals or to know what you're finding or not finding. Headphones also keep you from disturbing others. This is important if you're working a beach. With headphones on you can ease among the sunbathers and they'll hardly know you're there. Also, if you're in the southland, early-morning searching a beach in the winter, you won't want to disturb those fellows still wrapped in bedrolls.

Batteries are another essential need. The carbon-zinc batteries are fine, but alkaline batteries last longer. Buy a set and a spare. Never go into the field without extra batteries. You may think you just changed the batteries 'the other day,' but time slips by and you may learn to your chagrin that your batteries have expired.

In addition to detector, headphones, and batteries, you'll need a digging tool if you work on land, a light-weight plastic perforated scoop if you work in dry sand of a beach, and a long-handled steel scoop if you work in the surf. A probe is also essential — especially if you're working on land. Choose an ice-pick-like tool with a blunted end or one made of fiberglass that won't mar a coin as you try to locate it.

Perhaps you already have a sturdy garden trowel or hunting knife that will serve to dig your finds. And you may have a 2-pocket carpenter's apron for holding trash and treasure. For starters, use what you have at hand. After a few trips out, you'll realize what additional equipment would be nice to own.

Basic Field Tools for Treasure Hunters

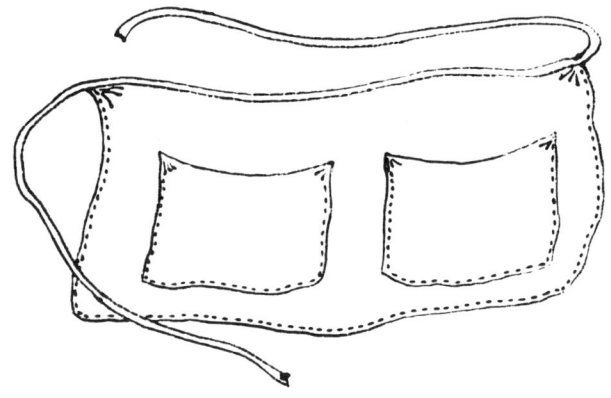

2-Pocket Carpenter's Apron

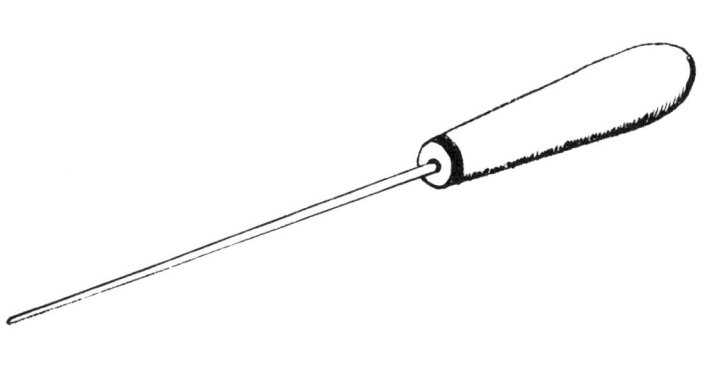

Probe / Ice Pick

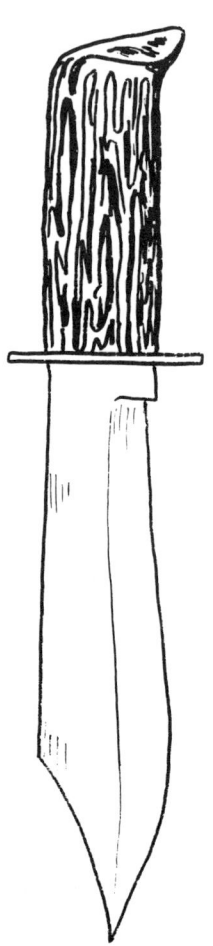

Hunting Knife

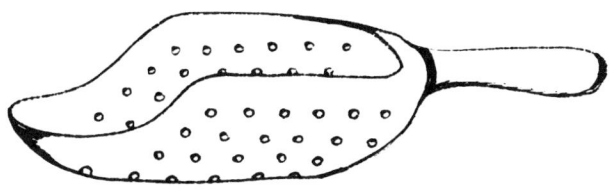

Perforated Scoop

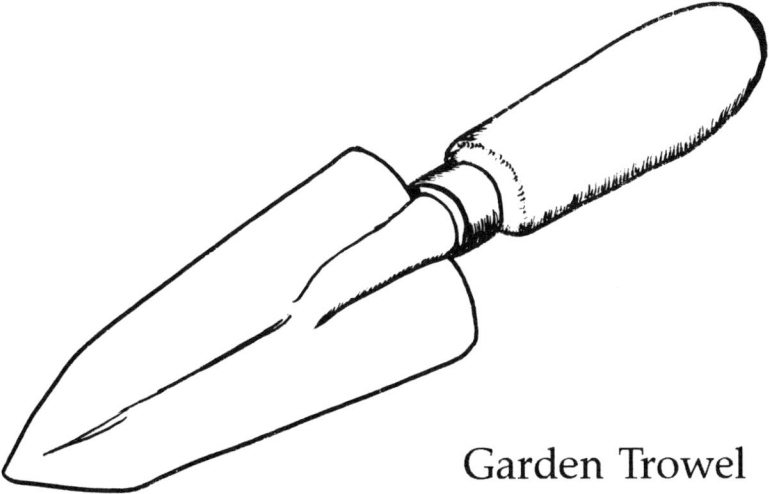

Garden Trowel

Get your equipment ready and I'll see you in the field.

Read Your Manual

Once you've received your detector, try to control your excitement. Don't destroy the box. Open it carefully and save it and the packing material. In the future you may want to use it as a travel case. Or you may need to return the machine for repair or modification or perhaps to trade it in on a more sophisticated model. Having the original box can save time.

Once you've taken the detector from its box, find the directions for assembling the parts and follow them carefully. This will be easy unless your hands are still shaking from the excitement of it all. Assembly usually consists of joining the two sections of rod stems, attaching the searchcoil to the stem with wing nuts, and attaching the searchcoil cable to the control box. The batteries will probably already be in place. But check on this. The machine won't work without batteries.

The next step is to read from start to finish the operating manual that's been included with the detector. I'm guessing that you're going to be so eager to try out the machine that you're going to skip-read this information. If you're too excited to read the whole manual, at least find the condensed operating instructions and read them carefully. And read a couple of pages to learn the purpose of the control knobs. You can return to the manual later for a more in-depth reading.

Instructions of the use of various types of detectors vary, so set the controls on your detector according to the instructions in your manual. Set the discrimination where it will eliminate nails and junk. Place your headphones on and plug them into the jack on the control housing.

Do as I say and not as I do. Wind the cable snugly on the detector stem. That loose cable will surely result in false signals as it moves about and acts as a target for my searchcoil.

You've already checked to see if the batteries are in place, but now it's time to check to see if the batteries are strong. Following your manual instructions, manipulate the control which provides battery information. If you get a strong signal, the batteries are in good shape. If the signal is weak, the batteries are weak. If there is no signal, get new batteries. It's a good plan to check the batteries each time <u>before</u> you go into the hunt field. If there's a problem, change the batteries at home. This will keep you from losing valuable search time in the field.

Be sure the cable is secured snugly to the rod stem so it doesn't touch the searchcoil and cause false signals. Hold your detector in a comfortable position a couple of feet in front of you and keep it in motion, swinging it slowly back and forth in a straight line. Hold it an inch or so above the ground, keeping the searchcoil parallel to the ground.

Work slowly, covering about 1 foot per second. Overlap your sweeps so you'll have a good chance of finding whatever's there to be found. (Nobody finds it all.) If you have an 8-inch searchcoil, overlap your sweeps by 4 inches. When you hear a beep you'll know you're near a target. That target should be directly below the center of your searchcoil. More about pinpointing the target later. For right now you're ready to go.

Before you venture widely into the field, which may be your back yard, you may want to place a few coins and other objects on the ground, pass your searchcoil over them, and listen for the sounds you'll receive. Careful listening will save you needless digging as you learn to operate your machine.

So practice by placing at two-foot intervals a penny, a nickel, a dime, a quarter, a nail, a pull-tab, a screwcap, listening carefully to the signals they cause your detector to emit. Modern dimes and quarters are called clad coins because they contain no silver. These coins will cause the detector to emit loud, harsh sounds which you will soon be able to recognize. Pennies give this same sound, too — especially pennies that are near the surface.

If you have some silver coins — coins minted before 1965 — place some samples on the ground and listen as the detector passes over them. You'll probably note that the sound is now more smooth and mellow.

Do you have some gold rings? If so, test your detector on a small yellow gold band, a larger gold band, a white gold ring. Listen for audio differences. Try to remember the sounds.

Now pocket your sample targets and try your machine in your yard. If the lawn is smooth and even, you may rest the searchcoil right on the grass. When you get a signal, run the searchcoil over it horizontally, then try it again running the coil vertically. You should get a smooth signal in both directions. If you receive a smooth signal in one direction but not the other, the target is apt to be junk. While you're learning to operate your machine, check out all signals just to be sure of how the detector is operating.

Once you hear a target signal, you need to pinpoint the target's exact location. It's sometimes hard to find a small coin within the area covered by your 8 or 10-inch searchcoil. If your detector has a pinpoint button, use it now.

Place your searchcoil on the ground slightly to one side of your target then push and hold the pinpoint button as you bring the coil back over the target area. Listen carefully. The spot where you hear the loudest signal should be the exact spot where you'll find your target. Locate the target with your probe.

You didn't find it? Read on.

If the coin target is lying flat in the ground, the pinpoint spot should be under the center of the searchcoil, but if the coin is lying on edge or at an angle, the pinpointing may vary and you'll have to search in a wider area. Probe carefully. You don't want to damage a valuable coin by scratching it.

As you work with your detector, you may become aware that the pinpointed target, even when lying flat in the ground, doesn't always fall exactly under the center of your searchcoil. Instruments vary. No two are exactly alike. To check for this variation, turn your detector on and run a coin over the bottom of the searchcoil. The spot where you receive the loudest signal is the bullseye spot on your detector. Mark that spot with a magic marker, top and bottom, and use it in the future.

Another way to pinpoint a target is to move the searchcoil left to right across the target area, noting where you get the loudest signal. Then move the searchcoil forward and backward, listening for the loudest signal. Under the spot where the two patterns cross giving the loudest signal, you'll find your target.

Still another method of pinpointing involves raising your searchcoil until you hear only the faintest signal. That signal should be directly over the target. Try these pinpointing systems until you find the one or ones that work best for you.

When I first began using my 1235-X, I had read that coins would cause the detector to emit a smooth single signal and that nails or elongated ferrous, or iron, objects would cause a double beep. I also heard the double beep on shallowly buried pulltabs. Easy, I thought. As long as I could count to 2, I had it made, didn't I?

Then I found my gold wedding band caused the detector to give a double beep. So did coins that were lying on edge. This seemed especially true for dimes. There were variables in this business! I felt I couldn't risk ignoring a double beep, so I was finding a lot of pull-tabs along with lots of coins.

By setting the discrimination at 7, I could avoid pull-tabs and that's what I did — for a while. With that setting I fould lots of pennies, dimes, quarters and very few pull-tabs. But where were the nickels? Had people stopped losing nickels? A more careful study of the manual told me that by setting the discrimination at 7 I was eliminating nickels and gold rings.

Finally, I developed a system of avoiding most pull-tabs that works for me most of the time. I set the discrimination at 4 because this should allow me to pick up gold rings and nickels. When I receive a signal, I then change the discrimination setting to 7. If the detector still beeps, I know the target isn't a nickel nor is it gold because those items would be discrimiated at 7. I know I have a pull-tab or a coin. Pushing the discrimination to 8 usually tells me which one because most pull-tabs are discrimated at 8.

The detector's sensitivity setting will play a part in helping you find deeply buried coins. Keep it set at around 8 unless you're receiving lots of false signals. If this happens, those signals can sometimes be eliminated by lowering the sensitivity setting. Realize, of course, that you lose detection depth when you lower that setting.

People will frequently ask how deep a detector will detect a coin. There are many variables. In general you'll find that the larger the object, the deeper the detector will detect. It's easier to find a quarter six inches deep than it is to find a dime at that depth. The dampness or dryness of the soil can also make a difference in depth detection.

Another favorite question of onlookers is: Have you found enough to pay for your detector yet? You don't have to give a direct answer, of course. If you've only been into the hobby a short time, the answer will probably be no. But if you keep swinging the searchcoil diligently, your answer may soon be yes. In fact, you may find one item that's worth far more than the price of your detector. I'm still waiting for this to happen.

How much did that thing cost? That's another question you'll hear frequently, especially from children and foreign tourists. Again, you needn't answer. In fact, by answering you may place yourself in jeopardy. Someone learning that your detector cost several hundred dollars may decide that they'd like to have it. Be guarded in your responses. When I'm asked that question, I say that I have no idea of the price since the detector was a gift.

After you've read this chapter, you may want to re-read your operating manual. It's a good plan to review it now and then. Information that wasn't meaningful to you a week ago may take on new importance in light of your continued detecting experience.

Retrieving Your Treasure

It's one thing to locate a target with your detector, and it's another thing to retrieve that target from the ground, the sand, or the sea bottom.

Practice searching and retrieving on your own property. Unless your yard borders a beach, you'll probably be working in soil and you'll have to dig in order to find your target. You must learn to do this without leaving your yard looking as if gophers have moved in.

First, pinpoint your target, then locate it with a probe. After you have probed gently and touched your target, note the depth of the probe. Remove the probe and insert and center an 8-inch screwdriver just above the target and rotate it slowly to open the ground. Then insert the screwdriver just under the target (remember that measured depth) at an angle, and lever the target to the surface. Tamp the earth back into place with your foot, leaving the ground as you found it. This system works well if the target is 1 to 4 inches deep and the ground is damp.

If the ground is hard and the target deep, use a sharp, sturdy hunting knife, cutting three sides of a 3 or 4-inch plug of soil and leaving the fourth side intact. Fold this plug of soil back. With your searchcoil relocate your target. If it's in the plug remove it with your fingers and replace the plug. If the target is still in the hole, dig it out with your trowel.

This may involve removing quite a bit of soil. Pile this soil on a cloth or a piece of plastic. After you've retrieved your target, funnel the soil back into the hole and replace the plug. Tamp on it with your foot. If you've worked carefully, it's unlikely that anyone will be able to tell where you've dug.

What did you find? Coin? Ring? Pull-tab? Whatever you find, it's a keeper. Your apron should have at least 2 pockets — one for trash and one for treasure. Removing the trash and disposing of it properly improves our environment, prevents you from finding the same trash next week, and makes one less piece of junk for another hobbyist to contend with.

In Iowa where I do a lot of my treasure hunting, the soil during mid-summer gets almost rock hard. Some treasure hunters recommend water hunting during these dry periods, but I've found that hard dry soil needn't completely defeat me.

While valuable silver coins may be deeply buried in inaccessible places during drought times, many modern coins are hidden only 1 or 2 inches deep. In fact, some are barely covered by a thin layer of grass or leaves. Such targets are easily found with a probe or sometimes by merely brushing the grass aside. When searching hard ground, I save time by setting the sensitivity low so I won't hear about the deeper targets that I'm unable to dig.

Don't give up during drought periods. Those clad coins are fun to find and they're still accepted in the shops. I've also found silver coins very near the surface — coins the oldtimers who searched the area 20 years ago must have missed. Remember: nobody finds them all.

If you want to search on private land, get permission from the property owner. If you want to search in public parks, get permission from the person in control of the area. Sometimes this can be as simple as talking with the caretaker. Other times it may require a visit or a letter to City Hall. Do it.

If you want to search in state parks, find out the rules and regulations covering those parks before you begin your search. Public beaches, unless they are in restrictive parks, are usually open to treasure hunters. Always avoid trespassing. By keeping your searches legal, you help develop good rapport between treasure hunters and the public.

If you're lucky enough to live near a beach, you'll probably enjoy working in the dry beach sand. There, it's a simple matter to retrieve treasure using a perforated sand scoop. The plastic models don't tire the hand as much as the steel ones, but the sturdy steel varieties allow you to work the heavy wet sand near the waterline where water action will wash sand through the perforations.

You needn't pinpoint as carefully when you work in dry sand because one deep dig will usually place the treasure in your scoop. If you don't retrieve your target on the first try, scoop again. After you've found your target, run your searchcoil over the hole again. You may get another signal. You don't want to miss that second one! This is good advice whether you're working in soil, or sand, or water. Always recheck the hole.

After you've retrieved your target and rechecked the hole, use that hole as a pivot point, sweeping your searchcoil out from it as spokes might radiate from a wheel, hub to rim. This is working on the theory that people seldom lose just one coin. There may be others nearby. Always refill holes whether you're working in soil or sand. You don't want someone to turn an ankle in a hole you've left.

If you're working in wet sand away from the water, a sand scoop will slow you down because it will take too long to shake it through the perforations. It's more practical to dig wet-sand targets with a trowel. Dig a trowel of sand, piling it to one side. Run your searchcoil over it, repeating the process until you get a signal. Then either locate the target with your fingers, or take time to work that small amount of sand through your scoop. Again, recheck and refill your hole.

If you're working in the shallow surf, from the water's edge to a depth of 4 or 5 feet, use your sturdy long-handled scoop. Here, pinpointing is very important. Once you've located your target, hold your searchcoil over it. Move your right foot until your toe touches the searchcoil, then move the searchcoil away and place the blade of your sandscoop a couple of inches behind your toe. Bend forward as you force the scoop into the sand with your foot, then straighten up and lift the scoop. As the scoop comes up, water action will wash the sand away, revealing your target. If you missed, scoop again. At first you may feel as if you need either 3 hands or a caddy, but practice will smooth your retrieval technique.

The following code of ethics is followed by most treasure hunters. You may wish to make it your own.

The Treasure Hunters Code of Ethics

I will respect private property and do no treasure hunting without the owner's permission.

I will fill all excavations.

I will appreciate and protect our heritage of natural resources, wildlike and private property.

I will use thoughfulness, consideration and courtesy at all times.

I will build fires in designated or safe places only.

I will leave gates as found.

I will not destroy property, buildings, or what is left of ghost towns and deserted structures.

I will not tamper with signs, structural facilities or equipment.

I will not litter.

Keep this code in mind as you pursue and enjoy your new hobby.

Treasure Awaits You

Let's consider treasure as anything of value to you or to others. By this time you've probably already decided what kind of treasure you'll search for. You may have good reason to search for gold nuggets or a cache of hidden loot. An Iowa man I know plans to search for modern arrowheads. He often takes his family camping in an area that bow-and-arrow hunters frequent during deer season. He spotted some lost arrows and he plans to search for more with the thought of reselling them.

However, most beginning detectorists will be interested in searching for coins and jewelry. The best advice I received when I began coinshooting was: look in your hometown first. Many times people are so familiar with their own towns that they can't imagine any kind of treasure awaiting them so close to home. But it does. Search locally. You'll probably find treasure only five or ten minutes from your doorstep.

Consider where treasure might have been lost or hidden. Wherever large groups of people have congregated, they will have left coins and jewelry behind. Make a list of such places in your town. If you've lived in that town a long time, consider your childhood. Where did people congregate then?

I can remember sitting on my grandparents' front porch and listening to a minister conduct a revival meeting in a tent on the vacant lot next door. Great crowds flocked in. I wonder what that lot holds today, and someday I'll go back and see, for surely there must be old coins hidden in that ground.

After you've listed some potential treasure spots, stop by your city hall or police station and see what restrictions, if any, might apply to metal detectors.

By now you've probably made quite a list of places to search. School grounds. Fairgrounds. Ball fields. City parks. Swimming pool areas. Courthouse grounds. YMCA/YWCA grounds. Private lawns. Churchyards. Amusement parks. Bandstands. Battle sites. Beaches. Benches. Bleachers. Boardwalks. Campgrounds. Circus and carnival lots. Concession stands. Drive-ins. Forts. Garages (especially those with dirt floors). Homesteads. Inns. Lakes. Libraries. Gravel or dirt parking lots. Parking meters set in grass. Race tracks. Revival meetings. Rodeo grounds. Service stations (old or new) with gravel driveways. Sidewalks (check the grass at the sides). Ticket booths. Vacant lots. Feel like getting started?

If there's a college in your town, it might be a good place to look for treasure, but ask permission first. Also, fraternity and sorority house lawns make good hunting grounds. These houses are usually privately owned, so even if the college grounds are off limits, you may be able to get permission to search the lawns of the Greek houses.

It's been my experience that if I approach a groundskeeper carrying only my detector and a slim probe, it's easy to get permission to search. Perhaps it's because I look less formidable than a person carrying detector, probe, knife, and trowel.

When I began searching my hometown parks, I had never seen other treasure hunters at work, and sometimes I found so many coins that I might have felt the area had never been searched, except for one thing. Almost all my coins were dated 1965 or later — the time era after the government had stopped minting silver coins. Sometimes I would find a few wheat-backed pennies which were minted until 1959, and this told me that there should be silver there, too. But I didn't find much.

Then I began to be approached by retired gentlemen and the conversation usually went something like this:

"I've got one of those machines at home in my closet," he said.

"Don't you use it anymore? I asked.

"Nope."

"Why not?"

"All the silver's been found and I'm not interested in those clad coins."

Then the man would tell me about all the Barber dimes, silver quarters, walking Liberty halves, and silver dollars he had found twenty years ago.

"This was a popular hobby twenty years ago," he said. "But most of us stored our detectors once the silver had been found."

I thanked him for his information as I continued to search the park. And I asked myself a few questions. Was the man right? Had all the silver been found? And had nothing of value been lost or hidden in the past 20 years? What kind of a detector had he been using? Had detectors not been improved in a score of years?

I continued to search. During my first year of searching I found a few silver coins — perhaps ones that kindly man had overlooked. But I didn't find many. Maybe I was missing the signals. Maybe I needed more practice in listening, more hours in the field.

Since I heard variations on this man's theme from many oldtime coinshooters, I feel that now is a great time to beginning this hobby. During those 20 years the likely spots for finding treasure have had time to fill up again. There is modern treasure to be found in addition to the silver coins that surely are still deep in the ground. Nobody finds it all.

As you gain more and more experience in coinshooting, you'll probably notice that you're finding a lot of pennies. Years ago, (Twenty? I found this information in a book bearing a 1970 copyright date), someone kept some statistics and came up with the information that 3 of every 4 lost coins were pennies. Why so many pennies! Pennies are small and they slip through the fingers easily. They have little worth, and a person realizing he has dropped one may just let it go. Lost coins of a larger denomination usually command a more careful search.

Today, the ratio of pennies to other coins found may be even greater than 3 out of 4. When I picked up a penny in a park, a mother nearby said, "My son throws pennies away."

"Doesn't he know a hundred of them make a dollar?" I asked.

"I keep telling him that, but he pays no attention. He pitches them."

A school custodian said, "Kids test their throwing ability with pennies. They throw them down the school hallways to see how far they can make them go. I sweep up a bunch of them almost every day. I just put them in a jar and about once a year I take my wife out to dinner."

So do as you will with pennies. Some people pick them up. Some people leave them in the ground. It might be a good club project to bring in the pennies and donate them to a charity. It would foster good public relations between treasure hunters and their community.

Search the spots close to home before you decide that the grass is greener in the next town or the next state. Search the easiest spots first, those with level mowed lawns and a smooth terrain. It's harder to work in weeds, rocks, or on hills.

Once you think you've run out of potential treasure places, do some research. Don't let the word scare you. Research can be as simple as asking an oldtimer where he used to swim or picnic. Once he starts talking, you may get lots of interesting information. And he may refer you to some of his friends who have additional tales to tell.

Do some library research on days when you can't get into a hunt field. Tell your librarian what you're doing and ask her advice on locating promising places to search. Ask to see old newspapers. They'll tell you where people congregated for picnics, revival meetings, fairs and carnvals. Check out some books on city and country history.

A visit to your local historical museum may also put you in touch with potential treasure spots. If you're not already a member of your local historical society, join. You'll be helping your community with your support and you may be given special treatment when it comes to searching local historical grounds. Of course you'll share part of your finds with your museum.

The treasure maps you see advertised in the treasure magazines are mostly novelties. They're like my first metal detector — good only for hanging on the wall. The maps that can help you are the ones produced by the U.S. Geological Survey. They show minute details of what was formerly located on a specific plot of land. Compare the old maps with a modern map. If the old map shows where an old school house used to stand, you might go to that spot, expecting to find some interesting treasures. The site may now be on private property, so ask permission. The first step in securing such a map is to write to the U.S. Geological Survey, GSA Building, Washington, DC 20242. Request an index to topographical maps of your state.

Once you decide which map or maps you want to purchase, you may order them from a regional office of the Geological Survey.

Once you get deeply into research, you're probably no longer a beginner!

Now is a good time to mention log books. Buy a notebook big enough to allow you to record your finds, their location, the date. List the number and denominations of coins found as well as their condition. Some day you'll want to be able to say, 'I've already found over a thousand coins.'

It's also a good idea to list the number of hours you spent in the field. You can use this figure to tell you about when you need to think about replacing batteries. I didn't do this when I first began coinshooting and one day when my searchcoil was right over a visible penny and I was getting no sound through the headphones, I thought my detector was broken. A set of new batteries repaired it. I simply hadn't realized that I'd spent 40 or 50 hours in the field.

Use these ideas for starters as you begin hunting your treasure. Listen to those old men with their tales of yesterday. They're savvy and they can offer you some good ideas. You might even be able to persuade some of them to get their detectors out of drydock and join you.

But most important of all, remember that nobody ever finds it all. There's still treasure out there just waiting for your searchcoil to pass over it.

Treasure On Land

This chapter refers to treasure lost or hidden in soil. Retrieval will require digging, so have the proper tools at hand. After you've decided where to go, you'll need to decide on the best time to go. Sometimes this is decided for you by the restrictions of job or family activities.

If it's summertime, you may want to search in the cool of morning or evening. In early spring or late fall you may welcome that sun and want to search at midday. Many treasure hunters prefer early morning hours because they offer greater privacy.

Consider the weather. You'll know better than to search in a thunderstorm, but you may enjoy working in a light drizzle. Just be sure both you and your detector's control housing are protected from the rain. Few strollers venture out in foul weather and again, you'll have a greater degree of privacy.

If you do encounter people who want to chat, take a few minutes to talk with them, explaining what you're doing. If these people are children, you can probably talk to them as you continue to search. They'll be curious about what you're finding. They may want to help you dig, I advise against giving them pennies. Once you do this, they'll tell their friends and you'll be playing Pied Piper. If you go on about your business, they'll soon get bored and go on about theirs.

It's possible to use your metal detector in the snow, but most people don't. If something is lost in a snowdrift, it's sometimes possible to find it then run back inside to warm up.

Always dress comfortably when you go treasure hunting. Dress in layers so you can match weather changes. Slacks that have plenty of bending room are essential. Kneepads offer comfort while you dig your targets. A pair of cotton jersey gloves will keep your hands clean. Wear comfortable shoes. If the ground is muddy, wear old shoes or rubber boots. Comfort is the name of the game.

Take care of the environment as you search. Digging targets and leaving the terrain as you found it takes some practice. Do that practice on your own property until you become proficient.

Be on the lookout for ideas for additional places to search after you get in the field. You may go to a park with the intent of searching the play equipment areas, then a casual walk over a hill may reveal a soccer or valley ball court. Lucky you!

Volley ball courts are good places to search, especially if they are sand covered. If you don't have a sand scoop with you, use your trowel. Volley ball courts have the added advantage of being used primarily for playing volley ball. Kids may drop their pull-tabs in a spectator's area near the court, but I find few pull-tabs right on the playing court. Only once have I searched a volley ball court without finding either coins or jewelry. Search these courts with your discrimination setting very low. Using low discrimination, you just might find that gold coin or ring you've been expecting.

In searching in my home town, I experienced a bit of serendipity at the state veteran's home. I had asked to search these grounds because the home is very old and would be a likely spot to find silver coins. Also, I knew that there had once been a lake there where the crowds picnicked and fished.

Due to a very wet spring, I had a hard time getting into this field. One day, as I stood looking at the former lake area, a caretaker on a mowing machine stopped to talk.

"Why don't you try that hill?" He pointed to a steep hill nearby. "Kids use that for sledding, and when I mow, I look down at lots of coins. Can't stop, though. Not on a hillside."

So I took his advice and picked up a handful of coins while I waited for the lake area to dry out. Remember that every area that gets snow has its 'killer hill' where the kids go to sled or ski. In fact, there may be more than one such hill. Observe them in winter. Hunt them in summer.

In talking to people, you may learn of additional search spots that are better than the places you originaly had in mind. They may tell you the former location of an old dump, or where someone is excavating for a new sidewalk or water line. Where dirt is being dug, you may find an interesting place to search. Coins that have been lost for years may turn up in the excavation for a house, basement, a driveway. Always search these spots if you can gain permission.

Picnic grounds are good places to go coinshooting. In decades past, before the advent of TV, picnics were a major form of entertainment. Today these old grounds yield coins. Art Schroeder of Brookfield, Illinois has made a hobby of finding coins dropped at old picnic grounds.

"Just look anywhere near a river," he says.

Art searched with his Fisher 1260-X in a picnic area that had been popular from the 1930's to the 1960's. In three month's time he found 137 coins including 15 Barber dimes, 7 Mercury dimes, 51 wheatback pennies, and about 40 Indian head pennies from the 19th century.

The really exciting news about these finds was that the 1916-D Mercury dime is worth $1,000 to $1,500 to coin collectors. But Art didn't sell. Instead he decided to try to collect the entire series of 77 Mercury dimes. He's coming close to realizing his goal.

When you first step onto a ball field or other large area that you want to search, you may wonder how to begin. How, in all that expanse of space, can you find the right spot to start searching with your 8 or 10-inch searchcoil? Some logical places come to mind. Check the most obvious places first — the area around a concession stand, areas under or around a boardwalk, the parking lot, the picnic area, the bleachers.

Once you've searched the 'hot spots' of an area, then consider the vast area of the field. If you want to search it all and have the time, begin in one corner, going from one end of the field to the other, being careful to overlap your sweeps.

Art Schroeder of Brookfield, Illinois has found thousands of coins at old picnic sites. This one, a 1916-D silver Mercury dime is worth over $1,000.

Once you reach the end of the field, turn, step a couple of paces to one side and make a return trip. Do this until the whole field has been checked. If you're not finding anything after several sweeps across and back, you may want to go on to a more promising spot. But if you decide to check the entire area, continue to work slowly and methodically. You may not be able to do this all in the same day. If not, mark your place and begin there on another outing. After you've checked the field moving north to south, recheck it moving east to west. You can also mark a large field into smaller segments and work a grid pattern over one segment before moving on to the next.

If you're in a big-space area that you won't be able to return to and you know there isn't time to check it all that day, try the X system. Imagine two diagonal lines that form an X as they cross the field. Begin searching on one of them, keeping your eye on the distant corner as you overlap your sweeps.

If you get a signal, dig that target and recheck the hole. Then using the target hole as your pivot point, search in sweeps radiating out from that hole. If you find no more targets, continue on your original line of search. If you do find more targets, you may have found a 'hot spot' on the field. Search it until it gives out. Once you've searched one entire diagonal line, search the other one.

You never can tell what's been lost or hidden beneath the land's surface. I'm still in awe of an instrument that can 'see' what I cannot see and help me find it.

Happy searching.

Treasure at the Beach

The day following a storm, Jim Owens had just begun searching a beach near Monterey, California when he found a gold coin. It was Christmas day and he was using a Fisher VLF-555D. The coin was the size of a nickel, and he was thrilled. Some Christmas present!

Later, after taking the coin to a dealer, he learned that it was a $5 variety on which the name of the Schultz & Co. had been misspelled. Jim Owens had found the 'big one.' Very few such coins were minted and he later sold his at auction for $45,000. The new owner believes it may eventually be worth $100,000.

Photo by V.M. Hanks, Jr. courtesy of Krause Publications.

Jim Owens of Ivanhoe, California found this extremely rare five dollar gold piece on a beach with his metal detector. He later auctioned it off in San Francisco for $45,000.

If that story isn't enough to make you grab your detector and head for the beach, consider this one.

Dick and Nancy Waters, while searching a beach in the Los Angeles area, found an extremely low tide which caused the stairs they usually used to get to the beach to hang about 6 feet off the sand. Using a Fisher 1210-X and a Fisher 1260-X, they found 1,764 coins in 21 days. It was during this hunt that Nancy found her first diamond ring, and following that she found a 20-inch 14K gold chain.

If you live near a beach, either fresh or salt water, your opportunities for finding treasure are magnified. Treasure will be found where people have been, and large numbers of our population frequent beaches. The next time you visit a beach, take time to observe, noting what people are wearing, what they are doing, and where they are doing it.

You'll see bejeweled rings, earrings, medallions, bracelets, chains, watches. People tend to place valuables on their blanket or beach towel while they swim. Then when they prepare to leave the beach, they may forget having done this and give their towel or blanket a shake. Too late they remember the jewelry.

Many people play rough and tumble games at the beach in which chains and wristbands are broken and coins are lost. Note where you see volleyball teams, frizbee throwers, kite flyers, touch football players. In those places you'll find treasure.

Some beaches are also good places to look for treasure that has been lost or hidden in ages past. If you live near a lighthouse, you can almost be sure that the structure was erected following disasters at sea. Sea captains might have been carrying gold and silver from the New World back to Spain when disaster hit. Their cargo may still be washing ashore.

Pirates seeking to safely hide their booty may have chosen to bury it on a beach which would be easy to find again later. Beaches near lighthouses are good treasure hunting grounds.

Your public librarian will probably be able to help you learn about historical beach sites in your area, and Coast Guard and Life Saving Service records can also alert you to likely treasure hunting spots.

Talk to the old-time residents of a beach area, asking where they swam years ago. Swimming sites change. If you can find the old beaches, you may be able to find treasure that other hobbyists have never thought of seeking. It's exciting fun to be the first one to search a treasure spot.

Remember that the sand on many beaches may be highly mineralized. Some metal detectors won't handle this well — especially in the wet sand areas. A VLF detector with the ability to discriminate will usually work well for you. Lowering the sensitivity setting sometimes helps avoid false signals.

Beach searching gives you many advantages in your search for treasure. You can go to the beach at any season of the year — as long as it isn't too cold. You can search at any time of day or night. But some times are better that others. Low tide is a good time because more sand is exposed to view.

Tide charts are available at most marinas and dive shops. The daily newspapers in beach areas usually carry tide information. Radio stations in many coastal cities have a weather station that gives continuous details concerning winds, weather, and tides. If you plan to search at low tide, find out the exact time of that tide, then plan your search accordingly so you're at the beach an hour or two both before and after the low ebb.

Wind conditions can play an important part in your decision of when to go to the beach. Offshore winds carry sand from the beach. This lowering of the sand thickness can bring hidden treasures within the depth of your searchcoil. Onshore winds can do just the opposite. They can pile sand up, placing hidden objects at depths inaccessible to you. But if there's an onshore wind, don't give up. Heavy objects may sink into deep sand, but lighter objects may lodge easily within detector range.

Storms cause a violent lashing of waves against shore. Heavy objects can be washed into shallow water. Plan to search a beach as soon after a storm as it's safe to do so. Whenever you beach search, try the easy places first. This is a fun hobby, not a marathon of physical endurance. Dry sand is easy to work. Wet sand is harder. Be kind to yourself.

Many times you'll have opportunity to do some research and to pick a beach that seems special to you, but many other times, you'll find yourself with only the opportunity of going to the nearest beach available. When you step onto a beach you may wonder how, in all that vast area, to choose a starting place. Be aware of certain things that might denote 'hot spots.'

If you see outcroppings of rocks or if there is driftwood lying about, search behind those things. Waves may have washed something valuable ashore and it may have lodged behind them. Search tidal pools carefully. Tidal pools or indentations where they have been mark spots where the sand layer is thinner and where heavy objects may have been trapped. I found a beautiful Italian sterling chain in such a pool on a Galveston beach.

After you've exhausted the easy and obvious search spots, you may want to begin at a hightide mark and search a straight line to the water. If the beach isn't crowded, place a marker at each spot where you've dug a target. After several down-and-backs you may see a pattern evolving.

You may have found a trough. Wave action sometimes leaves troughs in the sand, and in them objects heavier than sand may collect — shells, rocks, coins, jewelry. Once you've found a trough, follow it in a line parallel to the water until you've searched it thoroughly. After that, continue working paths perpendicular to the water, searching for another trough.

If you don't find another trough, mark an area and give it a grid-type search. I have worked an area in a grid pattern until I thought I had located every target, only to find more when I searched the same area, using diagonal lines.

Watch the action of the waves. At low tide you may see rivulets of water draining back into the sea. This marks a low spot in the sand. Where the sand is thin, you'll have a better chance of finding treasure. Sometimes these rivulets will make deep cuts in the sand. Check these out carefully and immediately because they can change or disappear quickly.

Sometimes wave action will leave large quantities of shells, gravel, rock and other debris. Check such spots carefully. There may be treasure hidden with the flotsam.

If you visit one special beach regularly, you may want to establish some water markers to use on future visits. If there is a permanent dock, try to leave a marker on it in a place that is slightly submerged during low tides. One man I know drives a nail into the dock piling to mark the tide line. A woman friend uses a nail polish marker. Now, when they visit that beach, they can tell at a glance how low the sand is. When they find lots of space between their marker and the sand, they begin seraching.

Many beaches will have piles of seaweed marking a high-tide line. This seaweed was once plants that grew just off shore and it was an important link in the food chain, feeding conch, sea turtles, sea urchins, and fish. The seaweed's usefulness continues after it's washed ashore. The dead plant material is food for other animals and it helps the beach and dunes perform their natural functions.

Perhaps seaweed hides treasure targets that might interest you. Search it with your detector. Some hobbyists rake the seaweed aside. If you intend to do this, check with the beach managers first to see if such action is allowed. In state parks that are being maintained as a natural setting, moving the seaweed may be restricted.

If nobody objects to your moving the seaweed and searching under it, move it toward shore so it can continue to serve its natural function of preventing sand erosion and fertilizing the sand dune plants. Many treasure hunters will not bother to search beneath the seaweed, so you may have first dibs on an untouched and promising location.

Beach areas are known to be full of trash as well as treasure, so be sure to wear your trash-and-treasure apron, saving the goodies and discarding the trash in refuse containers — preferably the ones at your home. If you rid yourself of the trash at the beach, you may do so hurriedly and thus discard something that deserved more study.

Metal detectorists need have little concern about damaging the beach. Don't disturb nesting birds, fill all your holes, and don't damage the sea oats. The grass-like sea oats is one of the few plants that can survive in moving sand that is dry and salty. These plants help hold the beach in place. If the sea oats die, the sand may be blown away and your beach will disappear. Because of the importance of sea oats, state laws may prohibit the picking of them, their flowers, their seeds.

Treasure hunting at a beach can spoil you for searching solid earth locations that require more strenuos digging. Take your detector to the beach every chance you get. Don't forget the spare batteries. You may stay longer than you planned to.

Treasure in the Surf

The bank vault of the sea and of fresh water lakes lies between the frothy shore line and water 4 to 5 feet deep. Beginners can easily search this treasure vault, but most of them don't. Where competition is the least, you may have the most luck, so why not put that searchcoil in the surf!

I'm still a beginner in surf searching, but as soon as I saw people working that area, I talked to them and a began reading on the subject. On a Kansas beach where my detector had remained silent, I began targeting coins as soon as I waded into the shallows. The same thing happened when I searched the surf area of a small lake near my home in Iowa.

Before surf hunting, be sure your detector is designed for that purpose. I didn't purchase a totally submersible detector at first because I wanted to see if I really liked surf hunting. I spent many hours using my 1235-X in the water. But I worried, knowing that if accidentally fell or if a wave caught me and wet down the control box which is not submersible, my search day would be ruined and my detector would need repair.

If you intend to water search with a land detector, be sure the searchcoil is submersible, not just splashproof or waterproof. Even if your coil is submersible, it's a good idea to seal the area where the cable connects to the coil with epoxy.

On most land machines, the control box isn't submersible. If the box is permanently attached to the detector, confine your searching to very shallow water. If the control housing is convertible, give it added protection by mounting it on your belt or on a neck and shoulder strap.

You may be able to use a short-handled retrieval scoop if you work in water no deeper than arm's length. In working at greater depths, you'll need a sturdy long-handled scoop.

When I first tried water searching, I found that my search-coil was so buoyant that I had trouble holding it near the bottom. I tried adding weights — rocks, sandbags, but I was spending more time getting ready to go than I was spending going. To solve the problem, I purchased an additional search-coil with a more open configuration.

Photo by Richard Francis

On a Florida Keys beach where my detector remained silent, I scooped up a watch (still ticking) after digging only a few shallow surf targets. A similar thing happened in a lake near my Iowa home. The dry beach seemed barren, but I began targeting coins when I stepped into the water. Note the sturdy long-handled scoop which I need when I'm working deeper water.

If you're searching in the surf, you'll need a trash-treasure apron with a secure closing — zipper, velcro, snap. You don't want water action to reclaim your goodies, nor do you want trash falling into the water where you may detect it again.

Water searching poses some retrieval problems. Pinpointing your target is very important because you're working only by touch and sound. At first you may feel you need more hands. If you're lucky you may capture your target in the first scoop, but don't be upset if it takes several tries. The retrieval process will become easier with practice. As a general rule, the shallower the water, the easier the retrieval. Be kind to yourself. Begin searching in very shallow water and, as your retrieval skill develops, gradually work to greater depths.

In surf searching, use as little discrimination as possible. A setting of 2 may enable you to miss most of the junk iron. Gold rings, gold doubloons, and pull-tabs are close to the same circuitry, so to discriminate the tabs may also mean to miss some gold. Don't take that chance. Keep the discrimination setting low and dig all signals. Seek out the easy digging spot first — soft sand. Save the hard mud, deep silt, or rocky bottom areas until you've mastered the soft sand techniques.

Earlier it was mentioned that post-storm days are good times to search the beach and surf areas. A surf hunter in the Keys related one of his post-storm experiences.

"It had been raining and blowing for several days, but on a Monday morning the rain had diminished to a gentle sprinkle. I took two hours off from work and my wife and I drove to the beach. She waited in the car while I splashed into the shallow water with my 1280-X. The surf was still rough, and where it beat against a bank, I could see gold rings tumbling in the current. I reached in with my hands and caught several, but more washed back out to sea before I could catch them."

Don't overlook shallow-water searching in your quest for treasure. It's out there for the finding.

Prospecting For Treasure

Using a metal detector, Bud Guthrie of Helena, Montana discovered one of the largest gold nuggets ever uncovered with a metal detector in the United States. In early February of 1990 while prospecting with a Fisher Gold Bug 'somewhere north of Phoenix,' he located and retrieved a fist-sized chunk of quartz which contained 60 troy ounces of gold. The value of his find is estimated at $75,000.

Bud Guthrie of Helena, Montana has found thousands of gold nuggets with his metal detector. This beautiful specimen is a chunk of granite laced with five pounds of gold.

Guthrie says he intends to keep the nugget, but that he might entertain serious offers for it because he makes his living by selling metal detectors and the nuggets he finds.

Some newcomers to treasure hunting, having read of experiences similar to Bud Guthrie's, come to the treasure hunting hobby with the express desire to search for gold. Others may not have prospecting in mind at first, but as they get more deeply into metal detecting and find they will be visiting an area where finding gold is a strong possibility, they, too, become interested in searching for nuggets.

If you have an all-purpose detector, you can give it a try in a gold-producing area, but you'll usually have better luck if you work with a machine that's been specially designed to find gold, such as Fisher's Gold Bug.

Nugget hunting differs from coinshooting. Coins are large compared to most nuggets. A one grain bit of gold is smaller than a match head and it will cause a very light signal. The nugget hunter must listen for faint sounds caused by very small objects and he must train his ears for those sounds.

The soil where gold is found is usually highly mineralized, so it is essential to follow the operating manual directions for tuning and ground balancing your machine. As you search you may visualize large nuggets, but you're more likely to find one of the smaller bits of gold. Use no discrimination as it will diminish the detector's sensitivity to minute nuggets. Dig every signal. You don't want to miss anything.

Always use headphones. Headphones allow you to hear faint signals that you will miss otherwise. Even a slight breeze can mask a nugget signal unless you're using headphones.

Keep the searchcoil as close to the ground as possible just as you do in most kinds of searching. For every inch you hold the coil above the ground you lose an inch of detector depth. Keep the searchcoil parallel to the ground, taking care not to swing it in an upward curve at the end of your sweeps. Overlap your sweeps without fail.

For the novice, the first few days of nugget searching may be frustrating. Hot rocks will sound just like nuggets, but you can learn to identify them without having to dig.

If you get a doubtful signal, sweep your coil over the spot from another direction. The signal should sound good from all directions. If it doesn't, you haven't found a nugget.

Another system of identification is to listen to the signal then raise the searchcoil a couple of inches. If you've found metal, the signal will become fainter as you raise the coil. With a non-metallic object the sound will quickly disappear.

Once you've targeted a nugget, don't let it get away from you. You may have to loosen the soil beneath your searchcoil with a pick. Then lift up a handful of the dirt. Run the searchcoil back over the targeted spot. If you still get a signal, grab another handful of dirt and check the area again. Repeat this action until the signal disappears.

At that time you'll know you have the target in your hand. Now sift a bit of the hand-held soil onto the top of your searchcoil. The top is as sensitive as the bottom. If there's no signal, shake that dirt away and sift more onto the coil. Continue this until you get a signal. At that time you'll know the nugget is on top of the coil. Pour this material into your hand and *gently* blow the dirt away. The nugget should remain.

In searching for gold, look in places where gold has been found in the past. The early miners discarded some gold in their tailings, especially in regions where gold was recovered with drywashers. Many areas are said to be worked out, but don't believe it. Nobody ever finds it all. Each season, wind, rain and flash floods expose new nuggets. Learn to use your machine. Search slowly and diligently. You're on the path to success.

Safety Tips

In pursuing any hobby there are common sense safety tips that it's wise to observe and this is true of treasure hunting. Do carry a waterproof first aid kit that's easy to open. Study its contents before you need them, so you'll be prepared to use them in an emergency. Be sure that kit includes your name, address, and telephone number as well as your doctor's name and number.

Wear suitable clothing that will keep you warm on cool days and cool on warm days. Beach and surf areas are full of broken glass and sharp metal. Always wear protective shoes both in and out of the water.

Wear a sunscreen with an SPF number of 15 or higher and reapply it often if you're in the sunshine and if you're in and out of the water. Sun rays can damage even on a cloudy day.

Stop hunting before you're exhausted. Sometimes you'll get so caught up in what you're finding or seeking that you may not realize you're expending too much energy. Don't become overtired.

If you're working in the sun and heat, keep a canteen of water with you to prevent dehydration. Take a drink frequently.

Keep physically fit. When you're working in an isolated or unfamiliar area, try to go with a partner. If this isn't possible, be alert to your surroundings and to danger. I don't view every stranger who approaches me as a potential serial killer, but that stranger might be surprised to know that the right hand in my pocket is gripping a vial of mace.

If you were ever a boy scout or girl scout as a child, you'll remember the motto — BE PREPARED. It's an excellent motto for treasure hunters, too.

Finders Keepers?

Who owns treasure? In the areas where you search, there may be local, state, and federal laws that apply to found articles. Learn what those laws are. Remember that all land and all property is owned by someone. In America, the laws concerning treasure trove encompass several points:

- Property intentionally abandoned by its owner belongs to the finder.
- Lost property still belongs to the owner. The finder should make every effort to find the owner.
- If property has been mislaid, misplaced, or hidden with no intention of abandonment, it still belongs to the owner. Law requires the finder to make a reasonable effort to locate that person.
- If the owner cannot be found, the treasure belongs to the owner of the land on which it was found, though the courts usually award a percentage of its value to the finder. Most people dislike being involved in a court dispute, so it's a good idea to have a written agreement with the property owner when you search on private land.

In trying to locate the owner of lost property, one can notify the police, place ads in a lost-and-found newspaper column, and use word of mouth advertising.

By writing to a nearby school and reporting the initials on a school medallion, I was able to return that bit of jewelry to its owner — or to the mother of its owner who was really pleased to have it back (because she's the one who probably paid for it). I never did hear from the boy whose ex-girlfriend actually lost the medallion on her local volleyball court.

What can you do with the treasure you find? You can keep it and enjoy it, of course. If you have a lot of valuable pieces, photograph them and use the pictures as your show-and-tell props when you share your finds with friends and family. Place the real treasure in a bankbox. If word gets around that you keep valuable coins or jewelry around the house, your house might become the target of burglars.

Perhaps you want to sell your finds. If so, be sure you know what you're selling. It's a good idea to have any jewelry or collector-type coins, buttons, or bottles appraised so you'll know their approximate value.

If you sell your finds to a dealer, you can expect to get about a third to half of the appraised value because the dealer must set a price that allows him to make a profit when he sells to his customers. In order to get a higher price, you can try to sell directly to the consumer through newspaper or magazine ads or through word-of-mouth advertising.

A man in Oklahoma who found a diamond ring appraised at over $3000 was pleased to be able to sell it directly to an acquaintance.

"It was a beautiful ring," he said, "and it fit me exactly."

"Didn't you hate selling it?" I asked.

He shrugged. "Yes, but my family really enjoys our new boat."

As you become involved in treasure hunting, don't forget the IRS. Treasure is nontaxable until you sell it. Once you sell it, you must report the sale as income. If you report treasure income, then you can usually deduct the expenses you incurred in finding that treasure — equipment, mileage, food, lodging, etc.

Treasure hunting is not always a case of finders-keepers, so protect yourself by knowing the law.

BIBLIOGRAPHY AND RESOURCES

MAGAZINES OF INTEREST TO TREASURE HUNTERS

"Treasure," Double Eagle Publishing Co., 31970 Yucaipa Blvd., Yucaipa, CA 92399

"Lost Treasure," Lost Treasure, Inc., P.O. Box 1589, Grove, OK 74344

"Western and Eastern Treasures," People's Publishing Co., P.O. Box 1095, Arcata, CA 95521

"World Treasure News," Fisher Reseasch Laboratory, 200 W. Willmott Rd., Los Banos, CA 93635

"Treasure Facts," Lost Treasure, Inc., P.O. Box 1589, Grove, OK 74344

VIDEO OF INTEREST TO TREASURE HUNTERS

"How To Get The Most Out Of Your Fisher 1200-X Series Metal Detector," with Joe Henderson, Host of South Carolina's TV show, "The Treasure Hunter." Fisher Research Laboratory, 200 W. Willmott Rd., Los Banos, CA 93635

BOOKS OF INTEREST TO TREASURE HUNTERS

Successful Nugget Hunting, Vol. I, by Pieter Heydelaar.
Advanced Nugget Hunting with a Fisher Gold Bug Metal Detector, by Pieter Heydelaar & David Johnson.
Advanced Shallow Water Treasure Hunting with a 1280-X Aquanaut, by Wallace L. Chandler.
America's Lost Treasurers, by Michael Paul Hensen
Coinshooting II, Digging Deeper Coins, by H. Glenn Carson

The books listed above are available from:
Fisher Research Laboratory
200 W. Willmott Road   Dept. MDT
Los Banos, CA 93635
Phone (209) 826-3292

There are books, pamphlets, catalogs, and newsletters available to those of you wishing to get more deeply involved in the hobby of treasure hunting with metal detectors.

If you would like information on how to receive any of the materials mentioned above, please contact:

Mr. Jim Lewellen, President
Fisher Research Laboratory
200 West Willmott Road
Attn: Dept. MDT
Los Banos, CA   93635

FIELD NOTES

FIELD NOTES

FIELD NOTES

FIELD NOTES

FIELD NOTES

FIELD NOTES

FIELD NOTES

FIELD NOTES